Shamsuddeen Umar

HACCP During the Production of Traditional Meat Snacks

AF138646

Shamsuddeen Umar

HACCP During the Production of Traditional Meat Snacks

Microbiological hazard and critical control point analysis of dried and minced meat snacks produced in Kano Nigeria

LAP LAMBERT Academic Publishing

Impressum / Imprint

Bibliografische Information der Deutschen Nationalbibliothek: Die Deutsche Nationalbibliothek verzeichnet diese Publikation in der Deutschen Nationalbibliografie; detaillierte bibliografische Daten sind im Internet über http://dnb.d-nb.de abrufbar.

Bibliographic information published by the Deutsche Nationalbibliothek: The Deutsche Nationalbibliothek lists this publication in the Deutsche Nationalbibliografie; detailed bibliographic data are available in the Internet at http://dnb.d-nb.de.

Coverbild / Cover image: www.ingimage.com

Verlag / Publisher:
LAP LAMBERT Academic Publishing
ist ein Imprint der / is a trademark of
OmniScriptum GmbH & Co. KG
Heinrich-Böcking-Str. 6-8, 66121 Saarbrücken, Deutschland / Germany
Email: info@lap-publishing.com

Herstellung: siehe letzte Seite /
Printed at: see last page
ISBN: 978-3-659-56107-8

Zugl. / Approved by: Bayero University, Kano 2009.

TABLE OF CONTENTS

LIST OF TABLES   Pages

LIST OF FIGURES Pages

LIST OF PLATES

LIST OF APPENDICES   Pages

ACKNOWLEDGEMENT

Praise be to God the Almighty. I want to express my sincere and profound gratitude to Professor J. B. Ameh, of Ahmadu Bello University Zaria and Professor T. I. Oyeyi of Bayero University Kano, for their guidance, support and advice.

I am also grateful to my friends and colleagues, Dr. Muhammad Yusha'u, Dr. Aisha Dantata, Dr. Hajara Haruna and other people like Tracy Efecodo, Zakari N. Lambu, Ibrahim G. Fagge, Adnan I. Jibril, Muhammad Salisu, Abubakar Y. isma'il and Abdurrazak M. Idris for their contributions.

CHAPTER ONE

1.10 INTRODUCTION

Kilishi (dried meat snack) is a dried and roasted meat snack traditional produced in Northern part of Nigeria and is consumed within and outside the country. *Dambun nama* (minced meat snack) on the other hand is a minced meat snack also traditionally produced by kilshi producers and is also patronized within and outside Nigeria.

Originally meat is a term used to describe any solid food, but has now come to be applied almost solely to animal flesh. As such it has played a significant role in the human diet since the days of hunting and gathering (Adam and Moss 1995).

Raw meat remains an important and probably the major source of human food borne infection with pathogenic bacteria. In spite of decades of efforts it has been difficult to obtain food animals free of pathogenic bacteria. Cattle and poultry carcasses are frequently contaminated with pathogens from the intestinal waste and from fecal material in the slaughter area (Dincer and Baysal, 2004).

The marketing of raw and processed meat products (*suya, balangu, kilishi* and *kanda*) is very popular in Northern Nigeria and the popularity of *suya* has extended to the Southern parts of the country (Umoh, 2001). According to Okonkwo *et. al.*, (1994) Nigeria and most tropical countries are said to have large markets in processed meat products. The diversity of ready to eat meat products is extensive, as they vary from one country to another as well as from one vendor to another. *Tsire, balangu, kilishi* and *dambun nama* are among the most popularly produced and consumed traditionally prepared meat products, especially in Kano.

Due to the putrefactive activity of certain bacterial enzyme, meat and meat products are highly perishable. According to Forest *et. al.*, (1975) meat and meat products are extremely perishable, so special care and handling must be exercised during all operations. It is also necessary to minimize deterioration in order to prolong the time within which an acceptable level of quality is maintained.

Even though there are no surveillance activities in Nigeria, laboratory studies have shown the presence of high microbial load as well as some organisms that are potentially pathogenic in street vended foods (Umoh and Odoba, 1999). Also studies on the microbiological quality of some meat products such as *tsire* and *kilishi* have shown them to have organisms of public health concern (Igene, *et. al.*, 1989). All these are due in part to lack of knowledge of the concept of Hazard analysis and critical control point (HACCP) that would help to ensure safety and quality of the meat products.

Hazard analysis critical control point (HACCP) concept if applied could help to control or eliminate all sources of possible contamination in order to ensure the safety of the meat or any other food. This HACCP concept is simply a management system in which food safety is addressed through the analysis and control of biological, physical and chemical hazards from raw material, production, procurement and handling, to manufacturing, distribution and consumption of the finished products (NACMCF, 1997). The implementation of HACCP concept will help meat producers to study their production process and find, monitor and control at the critical points. Improvements in meat processing and equipmentmaintenace will help to control product safety (Dincer and Baysal, 2004).

1.20 JUSTIFICATION OF THE STUDY

The research is important particularly from the public health perspective. This is because there are thousands of producers of these meat products in Kano and millions of urban dwellers consume them daily. There seem to be no standardization and quality control in both processing and product delivery. This apparent lack of quality control predisposes the consumers to serious health risk. In a country where the health care system is in deplorable state coupled with wide spread poverty the public health implication of these fast foods is enormous because diseases communicated through them are rarely understood and poorly treated.

Also records showed the presence of high microbial load in processed meat products. The question of where does the serious contamination come from arose. For this reason, fresh meat at the abattoir, the floor of the abattoir, the water used in washing of the meat, the hands of butchers, the knives, the tables on which the meat is being kept and the processed meat need to be microbiologically assessed.

The final processed products (*kilishi* and *danbun nama*) need much to be investigated since it is possible that microorganisms can equally contaminate them during production or serving.

Also studies which attempt to compare the microbial load of the different products are not so common; this work will therefore help to investigate the variability in the microbial load of the different products.

Spice ingredients are thought to have some antimicrobial activities, and yet meat treated with spices have high microbial load. There is there fore the need to investigate the activities of extracts from the various spice ingredients, on bacteria isolated from the meat products.

2

1.30 AIM OF THE RESEARCH

The aim of this study is to assess the microbiological hazard and critical control point of some meat products processed in Kano metropolis.

1.40 OBJECTIVES

The specific objectives are

1) To study the microbial flora associated with *kilishi* and *dambun nama* sold in Kano metropolis.

2) To carry out a study on the step by step processes adopted by the meat processors using HACCP concept in order to be able to identify the microbiological hazards and the critical control points.

3) To investigate the sensitivity of some of the bacteria isolated from the meat, to some of the common antimicrobial agents in use, and also their sensitivity to water and ethanolic extracts of the ingredients used in spicing the meat products.

CHAPTER TWO

2.10 LITERATURE REVIEW
2.1. 1. WHAT IS HACCP

Hazard Analysis and Critical Control Point (HACCP) is a preventive approach to consistent safe food production. It is based on two important concepts of safe food production; prevention and documentation (Norman and Robert 2006). HACCP therefore, is a management system in which food safety is addressed, through the analysis and control of biological, physical and chemical hazards from raw material production, procurement and handling, to manufacturing, distribution and consumption of the finished products (NACMCF, 1997).

The HACCP concept was developed in the 1950s through the National Aeronautics and Space Administration (NASA) and Natick laboratories for use in aerospace manufacturing under the name "Failure Mode Effect Analysis. This rational approach to process control for food products was developed jointly by the Pillsbury Company, NASA and the U.S. army Natick Natick Laboratories in 1971 in an attempt to apply a zero-defects program to the food processing industry (Norman and Robert 2006).. HACCP was incorporated to guarantee that food used in the U.S. space program would be 100% free of bacterial pathogens. Clark, (1991) described HACCP as a simple but a very specific method to identify hazards and for implementing the appropriate control to prevent potential hazards. Designed to prevent and not to detect food hazards, HACCP was identified by the U.S. Department of Agriculture (USDA) Food Safety and Inspection Service (FSIS) as a tool to prevent food safety hazards during meat and poultry production (Norman and Robert 2006).

2.1.2. Definition of terms.
Hazard

This is any biological, chemical or physical agent that is reasonably likely to cause illness or infection in the absence of its control.

Critical Control Point

This is a step at which control can be applied and is essential to prevent or eliminate a food safety hazard or reduce it to an acceptable level.

Critical Limit

This refers to a maximum and/or minimum value to which a biological, chemical or physical parameter must be controlled at a CCP to prevent, eliminate the occurrence of a food safety hazard or reduce it to an acceptable level.

Corrective action

This refers to the procedures followed when deviation occurs.

Deviation

This means failure to meet critical limit. (NACMCF, 1997).

2.1. 3. HACCP PRINCIPLES

HACCP is a systematic approach to the identification, evaluation, and control of food safety hazards based on the following principles (Norman and Robert 2006).:

  i. Conduct a hazard analysis.

- Identify steps in the process where the hazards of potential significance occur.
- List all identified hazards associated with each step.
- List preventive measures to control hazards.

  ii. Identification and documentation of the CCPs in the process.

  iii. Establishment of critical limits for preventive measures associated with each identified CCP.

  iv. Establishment of CCP-monitoring requirements, including monitoring frequency and person(s) responsible for the specific monitoring activities.

  v. Establishment of corrective action to be taken when monitoring reveals that a deviation from an established critical limit exists. The action should inclue the safe disposition of affected food and the correction of procedures or conditions that cause the out-of-control situation.

  vi. Establishment of procedures for verification that the HACCP system is working correctly. Responsible company personnel should conduct verification of compliance with the HACCP plan on a scheduled basis.

  vii. Establishment of effective record keeping procedures that document the HACCP system and update the HACCP plan when a change of products, manufacturing conditions, and evidence of new hazards occur (Norman and Robert 2006).

2.1. 4. Application of HACCP

For a successful implementation of HACCP plan, management must be committed to the HACCP concept. A firm commitment to HACCP by the top management provides company employees with a sense of importance of producing safe food (NACMCF, 1997).

One of the major reasons why hazards exist in many meat and meat products as well as other locally processed foods is lack of the application of such systems as HACCP which is a systematic approach to the identification, evaluation, and control of food safety hazards, based on the seven principles stated earlier. According to Umoh, (2001), in order to reduce contamination and therefore lower the risk of food born illnesses resulting from eating ready-to-eat meat product, meat inspectors and all involved with meat processing should be made to imbibe the HACCP concept.

To develop a better understanding of the microbiological problems associated with food production process, it is extremely necessary to apply the HACCP strategy (Oranusi *et. al.,* 2003). The HACCP strategy identifies hazards associated with different stages of preparation and handling assesses the relative risk and identifies points where control measure will be effective (Ehiri, *et. al.,* 2001). According to Bryan, (1992), experience has shown that HACCP system provides a greater assurance of food safety than other approaches such as the traditional quality control by testing the end product, and that the HACCP approach can be applied to food safety in homes and in food processing and in food service establishment.

2.1 .5 Method of identification of CCP

Critical control point is identified using the critical control point decision tree in Fig. 2.1. A step is a CCP if it involves a hazard of sufficient likelihood of occurrence and severity to warrant its control, control measure at the step exists and control at the step is necessary to prevent, eliminate or reduce the risk of the hazard to consumers.

CRITICAL CONTROL POINT DECISSION TREE

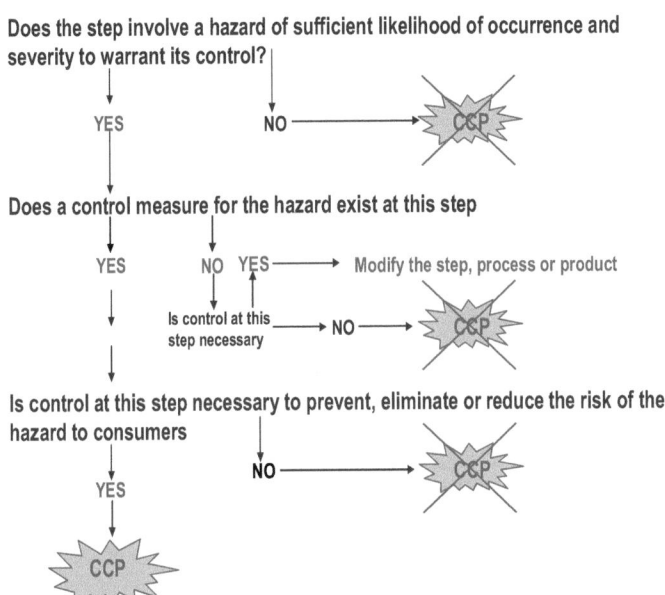

Source: NACMCF, (1997).

Fig. 2.1: Critical Control Point decision Tree.

2.20. Meat

Originally meat was a term used to describe any solid food, but has now come to be applied almost solely to animal flesh. As such, it has played a significant role on the human diet since the days of hunting and gathering. Though meat eating is abjured by some, on moral or religious ground, it remains widely popular today. In the main this is due to its desirable texture and flavour characteristics, in addition to the presence of protein of high biological value (Adams and Moss, 1995).

Forest *et. al.,* (1975), defined meat as those animal tissues, which are suitable for use as food. All processed or manufactured products which might be prepared from these tissues are included in this definition while nearly every species of animal can be used as meat. Meat is an excellent source of high quality protein, vitamins and certain minerals especially iron. It is easily digested and when cooked. Lean meat supplies nutrients which contribute significantly to the dietary balance of meals (Price and Schweigert, 1971). The chemical composition of a typical adult mammalian muscle after rigor mortis is as presented in (Table 1).

Table 2.1: Chemical composition of adult mammalian muscle.

Composition	Percent weight
Water	75.0
Protein	19.0
Lipid	2.50
Carbohydrate	1.20
Soluble non protein nitrogen	1.65
Inorganic materials	0.65

Source: Adams and Moss (1995).

2.21. SOME MICROORGANISMS OF PUBLIC HEALTH IMPORTANCE ASSOCIATED WITH MEAT AND MEAT PRODUCTS

Food and microorganisms have a long and interesting association that developed long before the beginning of recorded history. Foods are not only of nutritional value to those who consumed them, but often ideal culture media for microbial growth. Microorganisms that have been incriminated in food born illness resulting from consumption of meat and meat products include; *S. aureus, Salmonella spp, listeria monocytogens, Yersinia enterocolitica, Bacillus cereus, E. coli, Clostridium perfringens* as well as yeast and moulds (Saide-Alboronz *et. al.,* 1995).

2.2. 2. *Staphylococcus aureus*

This is a gram positive facultatively anaerobic prokaryote whose spherical cells are typically clustered in grape like arrangement. Staphylococci are non motile, salt tolerant and are capable of growing in media that are up to 10% NaCl which explains how they tolerate the salt deposited on human skin, by sweat glands (Robert, 2004). *Staphylococcus aureus* possess several toxins that contribute to their pathogenicity including cytolytic toxins, exfoliative toxins, toxic shock syndrome and enterotoxins (Robert, 2004). *Staph. aureus* is one of the more common causes of food poisoning. Commonly affected foods include processed meat, custard pastries, potato salad and ice cream that have been contaminated with the bacteria from human skin (Robert, 2004). For the bacteria to grow and secrete toxins, the food must remain at room temperature or warmer for several hours. Warming or reheating contaminated food does not inactivate enterotoxins, which are heat stable although heating does kill the bacteria. Food contaminated with staphylococci does not appear to taste unusual (Robert, 2004). Food poisoning by *Staph. aureus* is characterized by a short incubation period typically   two-four hours. Nausea, vomiting, stomach cramps, retching and prostration are the predominant symptoms although diarrhea is often reported and recovery is normally complete within one-two days. In severe cases dehydration, marked pallor and collapse may require treatment by intravenous infusion (Adams and Moss, 1995).

2.2. 3. *Escherichia coli*

It is a Gram negative, catalase positive, oxidase negative, fermentative, short non spore forming rod which ferments the sugar lactose. *E. coli* can be differentiated from other members of *Enterobacteriaceae* on the basis of a number of sugar fermentation and other biochemical tests. Classically an important group of tests used for this purpose are known by the acronym IMViC that is indole, methyl red, Voges proskaurer and citrate tests (Adams and Moss 1995). *E. coli* was identified as food pathogen acquired through the consumption of undercooked ground beef (Rajworsky and Marmer, 1995). Also according to Erickson *et al.,* (1995), *E. coli* has been identified as pathogen acquired through the consumption of undercooked ground beef. There are four categories of diarrhoeagenic *E. coli* based on distinct plasmid encoded virulence properties. They are ETEC, EIEC, EPEC and EHEC. EHEC O157:H7 is the one most frequently isolated from humans. It is more commonly involved in food borne transmission than other diarrhoeagenic *E. coli*. It also causes life threatening conditions, haemorrhagic colitis, haemolytic uremic syndrome and thrombotic thrombicytopaenic purpura (Adams and Moss,1995).

2.2. 4. *Salmonella.*

This is a motile, Gram negative, peritrichous bacillus that lives in the intestine of virtually all birds, reptiles, and mammals and is eliminated in their feces (Rebert, 2004). This bacterium does not ferment lactose, but it ferments glucose, usually with gas production. It is urease negative and oxidase negative, and most produce hydrogen sulfide (H_2S). Most infections of humans with *Salmonellae* results from the consumption of food contaminated with animal feces (Robert, 2004). *Salmonella spp.* Remains amongst the most important food borne pathogens worldwide. In recent years there has been a marked increase in food born salmonellosis with outbreaks being reported in several countries including Spain, Italy, England and America (Cloak *et. al.,* 1999). Contaminated food such as meat and meat products, that are in adequately cooked are the vehicles for the transmission of *Salmonella* spp which has been described as the leading cause of food borne bacterial disease (Nottermaus and Hoogenboom, 1992). Salmonellosis is the most frequently occurring bacterial food infection (WHO 1976, Gregory *et al.,* 1994). Humans are the sole hosts of *Salmonella typhi* which causes typhoid fever. Infection occurs via the ingestion of food or water contaminated with sewage containing bacteria from carriers, who are often asymptomatic. An infective dose of *S. typhi* is only about 1,000-10,000 cells (Robert, 2004).

2.2. 5. *Clostridium perfringens.*

This is an anaerobic, gram positive, endospore forming bacillus that is ubiquitous in soil, water, sewage and the gastro intestinal tracts of animals and humans (Robert, 2004). Although it is non motile, its rapid growth enables it to spread across the surface of laboratory media, resembling the growth of motile bacteria. *Clostridium perfringens* produces 11 toxins that lyse erythrocytes and leukocytes, increase vascular permeability, reduce blood pressure, and kill cells, resulting in irreversible damage (Robert, 2004). Clostridial food poisoning typically results from the ingestion of large members (10^8 or more) of type A *Clostridium perfringens* in contaminated meat (Robert, 2004). The food poisoning associated with *C. perfringens* is an infection rather than intoxication (Broidy and Kaminsky, 1978). *C. perfringens* is considered to be one of the three most important causes of food poisoning in the USA and UK (WHO, 1976).

2.2.6. Fungi and Mycotoxins.

The list of moulds associated with meat has been growing (Claude and Maurice, 1979). It includes fungi of the genus *Aspergillus, Penicillium, Mucor, Neurospora, Fusarium* and many others. On refrigerated meat, it is not uncommon to see white patches of *Sporotrichum carnis* or black patches of *Cladosporium herbarum* (Claude and Maurice, 1979). One important characteristic of fungi particularly the moulds is their ability to produce mycotoxins. Pitt, (1996), defined mycotoxins as fungal metabolites which when ingested, inhaled, or absorbed through the skin, cause lowered performance, sickness or even death in man and animals including birds. According to Miller, (1996), the single most effective and beneficial change, which could be made in human diet around the world, would be the elimination of mycotoxins.

2.30. MICROBIOLOGY OF MEAT AND MEAT PRODUCTS

Cattle production in Northern Nigeria, is mainly extensive, as most of the cattle are owned by nomadic herdsmen. Semi intensive system is practiced to some extent (where supplementary feed is given in addition to grazing). During movement of the animal in search of pasture and water, the herds come in close contact with one another especially at watering points thus increasing the risk of disease spread among them. Animals for slaughter are generally taken to cattle markets for sale. They may then be transported by rail, by trailers or on hoof to distant towns where they are slaughtered (Umoh, 2001). The animals may be infected in the posture, at the watering point, through the supplementary feeds, and during transportation. This infection may eventually be a source of contamination for the meat from the animal (Umoh, 2001).

Tissues of a healthy animal are protected against infection by a combination of physical barriers and the activity of the immune system. Consequently, internal organs and muscles of a freshly slaughtered carcass should be relatively free from microorganisms. Microbial numbers detected in aseptically sampled tissues are usually less than 10cfu/kg, although there is evidence that numbers can increase under conditions of stress and they will of course be higher if the animal is suffering from an infection (Adams and Moss, 1995). The most heavily colonized areas of the animal that may contaminate meat are the skin and gastrointestinal tract. Numbers and types of organisms carried at these sites will reflect both the animal's indigenous flora and its environment (Adams and Moss, 1995).

Meat is generally regarded as an excellent medium to support microbial growth due to its high water activity and abundant nutrients (Adams and Moss, 1995). Unless dried, cured, smoked, or fermented in some way, meat has a tendency to spoil rapidly (Robert, 2004). Meat and poultry are perishable foodstuffs, and red meat has a relatively unstable colour. Poor sanitary practices increase microbial damage resulting in reduced colour, flavour and product safety. Effective sanitation is essential to reduce discoloration, spoilage and pathogen growth with resultant increase in shelf life and product safety (Norman and Robert, 2006).

Several chemical factors are said to be involved in the stability of meat products, but the cheap concern for product stability is the control of microbial growth (CFIA, 1997). This control usually could be achieved by lowering the moisture content of meat to a critical level, thus inhibiting microbial growth (Fox and Ackerman, 1970). According to Bourquim (1980), meat preservation is possible without the use of chemicals by simply reducing the bulk of its water content. This drying of meat is a practice aimed at removing the bulk of the water and probably impart on the safety of the product (Abdullahi et al., 2004).

The drying condition of meat product must be sufficient to ensure rapid drying, there by reducing the time for microorganisms to grow (Holley, 1985, Abdullahi et. al. 2000). The microbial load of the product may hence be kept low if only good hygienic practices are followed at all stages of production (Frazier and Westhoff, 2006). Most food poisoning results from mishandling food, wrong temperature of storage, cross contamination and incorrect reheating.

2.3. 1. Spices used in the production of the various meat products, and their microbiology

During the production of the various meat products, spices are applied at one point or the other. These spices are normally used in powdered form in the case of and *dambun nama*. But in the case of *kilishi*, the spice ingredients are mixed in water, so that the dried meat is dipped into the liquid spice mixture, which normally consists of groundnut, ginger, pepper, cloves, West African black pepper, salt and seasoning.

Spices could be defined as the natural vegetable products or mixtures thereof, without any extraneous matter that is used for flavouring, seasoning and imparting aroma to foods (ISO, 1972). Spices like other food substances, may carry some bacteria, yeasts, moulds spores and even some insects. The predominant flora is generally composed of aerobic spore and non spore forming bacteria, indicator organisms and some pathogens may also be found (ICMSF, 1986). Coliforms were isolated and characterized to be *E. coli, Klebsiella spp, Pectobacterium spp* and *Enterobacter* (Patel *et. al.,* 1976).

According to Frazier and Westhoff (2006), spices do not have a marked bacteriostatic effect in the concentrations used in meat products and they may even serve as source of contamination of processed product. Occurrence of microorganisms that are potentially pathogenic in spices used in *suya* preparation is considered as major cause of gastrointestinal disturbances resulting from the consumption of suya in Nigeria (Ejeikwu and Ogbonna, 1998). According to Price and Schweigert, (1971), unless spices are treated to reduce their microbial content, they may add high numbers and undesirable kind of organisms to food in which they are used. The preservative effect of spices and condiments is attributed to the presence of some active antimicrobial agents contained in them, some of which are volatile oils, which posses' bacteriostatic properties. Resistance however varies with the microorganism and the spice in use. For the spice, it depends on its source, the freshness and the storage state (Frazier and Westhoff 2006).

2.3. 2. PRODUCTION OF *KILISHI* (DRIED MEAT SNACK)

The main source of all meat used in the production of the meat products is the Kano main abattoir, according to the producers. Beef and mutton could be used in the production, but most of the producers use beef.

The production of *kilishi* is a bit long process. It is accomplished by cutting the meat into thin flat sheets then sun drying it. This is followed by the application of spice, by dipping the dried meat into a spice mixture, which consists of groundnut, oil, pepper, cloves, onion, West African black pepper, seasoning and salt. After dipping, the meat is then sun dried again for the second time after which it is roasted on fire for about three to five minutes. After roasting, it is then piled in a wooden box with glassed sides for retailing.

2.3. 3. PRODUCTION OF *DAMBUN NAMA* (MINCED MEAT SNACK)

Dambun Nama on the other hand is produced by cooking of the raw meat in water with onion and seasoning until soft. The cooked meat is then minced using pestle and mortar, at this stage Pepper, seasoning and salt are also added after which it is fried in oil. The fried minced meat is then placed in a container with a perforated bottom for oil to drain out. It is then packaged in plastic containers for retailing.

CHAPTER THREE

3.1. 0. MATERIALS AND METHODS

3.1. 1. Sampling sites

Agadasawa was selected for *Kilishi* and *danbun nama*, All the selected areas are well known for the production and selling of the products sampled from there.

Agadasawa is located at main Kano city in Municipal local Government area behind Murtala Muhammad specialist hospital. It is a densely populated area and it is the largest *kilishi* and *dambun nama* producing area in Kano. The sides of the roads leading in to Agadasawa is populated with kiosks and tables of *kilishi* and *dambun nama* sellers. It is a typical Hausa community settlement. In addition to the *kilishi* sellers there are other people selling things like cola nut and provision stores.

3.1. 2. Sample collection and identification of microbiological hazards

The sample collection was done according to the method described by FAO (1979). A clean dry leak proof container (glass beaker) was used. Raw meat samples were collected from abattoir in the morning immediately after slaughtering of the animal. Same raw meat was also collected from *kilishi* and *dambun nama* producers before roasting. The spice and spiced meat (same batch) were also collected where applicable. The processed samples (*kilishi* and *dambun nama*) were collected from processors located at the different sampling points immediately after roasting and after holding for some time. In the case of *kilishi*, samples were collected the following day after the collection of the raw one, because it was not produced within one day. The samples were collected from, Agadasawa, popularly known for the production and sale of these meat snacks.

All the steps involved in the production of all the two meat products (*kilishi* and *dambun nama*) were carefully studied. This as achieved by staying with the producers during the production and giving attention to the temperature and time of holding and roasting. Bacterial and fungal populations were evaluated from utensils, knives, hands and then the spice, raw as well as the processed meat at various points of processing; microorganisms that are potentially dangerous were identified and recorded.

3.1. 3. Determination of the critical control points

All the steps involved in the production of all the two meat products (kilishi and dambun nama) were carefully studied. The results obtained from the microbiological analysis of each step was compared with the CCP decision tree (provided by U. S.

14

National Advisory Committee on Microbiological Criteria for foods NACMCF (1997), to establish whether that step is a critical control point or not. Below is the CCP decision tree.

CRITICAL CONTROL POINT DECISSION TREE

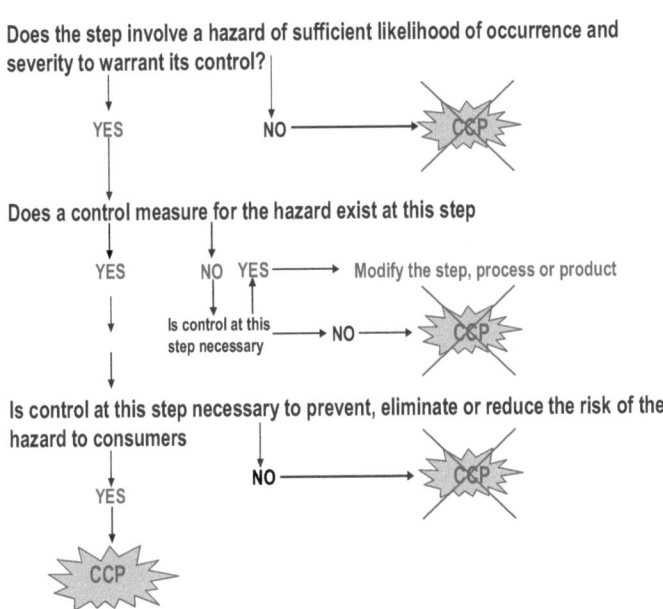

Source: NACMCF, (1997).

Fig. 3.1: Critical Control Point decision Tree.

3.2. 0. MICROBIOLOGICAL ANALYSIS

Microbiological analysis was carried out on the meat samples (raw and processed), the spices, the water and the swabs to enumerate, isolate and characterize some microorganisms of public health importance. (Uraih, 2004).

3.2. 1. Swabs

Sterile swab sticks were used to take swab samples from the floor of the abattoir, the knives of the butchers (raw and processed meat handlers), palms of the butchers and the tables on which meat is being kept. For the tables, floor and palms a $4cm^2$ cut was made on aluminum foil and this was used as the size of the area swabbed. For knives $1cm^2$ area was swabbed. The swabs were taken immediately to the laboratory for analysis. Four milliliters of peptone water was poured into the tube of the swab stick this was labeled as the stock homogenate. After mixing thoroughly, 1ml of peptone water from each of the tube was transferred to a test tube containing 9ml of peptone water. The tube was labeled 1:10 dilution. From this tube 1ml was transferred to another tube containing 9ml of peptone water and this tube was labeled 1:100 and so on up to $1:10^5$. From the homogenate a loop full of inoculum was streaked on to Eosine methylene blue agar for the detection of *E. coli*. Also from the serially diluted tubes 1ml of inoculum was transferred in to duplicate Petri dishes, this was followed by pouring of molten nutrient agar and malt extract agar for aerobic plate count and fungal count respectively. After incubation number of colonies was multiplied by dilution factor to get number of colony forming unit per cm^2.

3.2. 2. Sample Preparation and Serial Dilution

The sample preparation was carried out according to the method described by FAO (1979). In this method 25g of sample (meat and spice) was weighed and homogenized by blending in 225ml peptone water at 15,000-20,000 rpm. This was labeled as 1:10 dilution which is also the stock or the homogenate. This was further serially diluted to $1:10^7$.

3.2. 3. Total Aerobic Plate Count

This was carried out according to the method Adullahi and others, (2004). In this 1ml of inoculum from 10^3, 10^4, 10^5, 10^6 and 10^7 dilutions were transferred into duplicate Petri dishes which were labeled accordingly. This was followed by pouring aseptically about 15ml of molten plate count agar (nutrient agar). The culture was homogenized by swirling the plates and later allowed to solidify. The plates were incubated at 37°C for 24hrs. After incubation, plates containing 30-300 colonies were selected and the colonies counted. The average was taken and the number obtained was multiplied by the dilution factor. This gave the number of colony forming units per gram of a sample (cfu/g).

3.2. 4. Detection of *Staphylococcus*

Plates of mannitol salt agar were inoculated and incubated at 35°C for 24hrs. Following incubation, mannitol fermenting organisms which showed a yellow zone surrounding their growth were isolated on to agar slants for biochemical tests.

3.2. 5. Yeast and Mould Count

This was carried out according to the method of FAO (1979). In this 1ml of inocula from 10^3, 10^4, 10^5, 10^6 and 10^7 dilutions were transferred into duplicate petri dishes which were labeled accordingly. This was followed by pouring aseptically about 15-20ml of molten malt extract agar. The culture was mixed by swirling the plates and later allowed to solidify. The plates were incubated at 20-25°C for 2-5 days. After incubation, plates containing 30-300 colonies were selected and the colonies counted. The average was taken and the number obtained was multiplied by the dilution factor. This gave the number of colony forming units per gram of a sample (cfu/g). The fungi were identified using a colour guide and microscopy.

3.2. 6. Detection and Enumeration of Coliforms

This was carried out according to method described by Atlas, (1997). In this method, a set up consisting of 9 test tubes each containing 9ml of lactose broth and an inverted Durham tube, were autoclaved to expel air and to sterilize. Inoculation was made from the serially diluted samples as follows: From the 1:10 dilution, 1ml of inoculum was transferred to each of the first three of the 9 test tubes containing 9ml of lactose broth. Then 1ml also was transferred from 1:100 dilution to each of the second set of three test tubes of lactose broth and finally 1ml of inoculum was transferred from 1:1000 dilution to each of the last three tubes. All the 9 test tubes were incubated at 37°C for 24 hours and another 24 hours in the absence of gas (presumptive test). Following 24 hrs of incubation the tubes were observed for gas production and the number of gas positive tubes was compared with the most probable number (MPN) table to estimate the most probable number of coliforms per gram of sample. A loop full of inoculum from gas positive tubes was streaked on to

Eosine methylene blue (EMB) agar plate and incubated at 37°C for 24 hrs. Following incubation, colonies which formed bluish black colour with green metallic sheen, and reddish colonies were noted and isolated on agar slants. This is called the confirmatory test. Also colonies showing metallic sheen on EMB, were sub cultured in to tubes of lactose broth and incubated at 37°C. The tubes were observed after 24hrs., for gas production. This is the completed test for fecal coliforms. Colonies that formed green metallic sheen on EMB were stored on nutrient agar slant for further analysis.

3.2. 7. Detection of *E. coli* O157:H7

Isolates that formed green metallic sheen on EMB were streaked on Sorbitol Mackonkey Agar, and incubated at 37°C for 24 hrs. After incubation the plates were observed for the presence of colorless colonies. These were then tested using serological kit for the confirmation of *E. coli* O157:H7.

3.2. 8. Detection of *Clostrdium perfringens*

This was carried out according to method described by Cheesbrough (2000). The homogenate was streaked on to neomycin blood agar and incubated anaerobically at 37°C for 24 hours. Following anaerobic incubation, the plates were checked for large β-haemolytic colonies. The colonies were stored anaerobically on agar slants for further analysis.

3.2. 9. Detection of *Salmonella*

This was carried out according to FAO (1979). The homogenate was incubated at 37°C for 16-20hrs. After incubation, 10ml was transferred into 100ml selenite cystine medium and incubated at 37°C for 24 hours (enrichment). A loopfull from the enrichment medium was streaked on to brilliant green agar plates and incubated at 37°C for 24-48hrs. The colonies were gram-stained and tested for motility. The gram negative motile rods were inoculated on to nutrient agar slant for subsequent biochemical tests.

3.3.0. BIOCHEMICAL TESTS FOR THE CHARACTERIZATION OF THE MICROORGANISMS ISOLATED FROM THE SAMPLES.

3.3. 1. Motility Test.

This was carried out according to the method of Cheesbrough, (2000). Small drop of colony suspension was placed on a slide (the one with depression at the centre), and covered with cover slip. This was sealed with molten petroleum jelly and examined with 10X and 40X objectives for the presence of motile bacteria.

3.3. 2. Indole Test

This was carried out according to Cheesbrough, (2000). The test organism was inoculated into bijou bottle containing 3ml of sterile tryptone water, incubated at 35-37°C for up to 48hrs. This was followed by the addition of 0.5ml Kovac's reagent. Red color on the surface layer within 10 minutes indicated positive test for indole.

3.3. 3. Methyl Red Test

This was carried out according to Fawole and Oso, (2001). Two different tubes of MR-VP broth were inoculated with the suspected *E. coli* colonies. These were incubated at 37°C for 2-3 days. Five drops of methyl red indicator were added to each tube. Red color gave positive (acid) test and yellow color indicated negative (alkaline) test.

3.3. 4. Kligler iron agar test.

This was carried out according to the methods of Cheesebrough, (2000). Kligler iron agar (KIA) tubes were sloped and inoculated with the test organism from a nutrient slant. The butt was stabbed and the slope streaked, and incubated at 37°C for 24hrs. A positive result was indicated by a crack in the medium due to gas production from glucose, blackening of medium due to hydrogen sulphide (H_2S) production.

3.3. 5. Coagulase Test

This was carried out according to the method described by Cheesbrough (2000). A drop of distilled water was placed on each end of a slide. The test organism was emulsified to make a thick suspension. A loopful of plasma was added to one of the suspensions. This was mixed gently and observed for clumping. The other suspension was a control. Clumping within 10 seconds indicated the presence of *S. aureus*.

3.3. 6. Catalase Test

This was carried out according to Cheesbrough, 2000, Cappuccino and Sherman (2002). Hydrogen peroxide (2-3ml) was poured in a test tube. Some colonies of the test organism were removed and immersed using a glass rod into the H_2O_2 solution. After immersion of colonies it was observed for immediate bubbling.

3.3.7. Identification of moulds.

Identification was carried out using a colour guide (Frey *et. al.,* 1981) and staining using lactophenol cotton blue. The mycelium to be stained was transferd on to a drop of cotton blue in lactophenol on a clean glass slide. This was then observed under low power and high power of microscope (Fawole and Oso, 2001).

3.4.1. Hand culture experiment

Two Plates of nutrient agar were taken to one of the meat processors. He was asked to touch the surface of the agar with his hand before washing of the hand. He was then asked to wash his hand with klin detergent with bottle water (swan). The plates were immediately taken to incubator, and incubated at 37°C for 24 hours.

3.4. 2. Determination of pH of the different products.

The pH of the different products was determined using a pH meter by homogenizing the product in sterile deionized water, and dipping the electrode of the pH meter.

3.5. 0. PROXIMATE ANALYSIS

Proximate analysis of the meat products was carried out according to the methods of Egan *et. al.,* (1981).

3.5. 1. Total Ash determination

A clean dry crucible was weighed (W1). Five gram of the dried sample was placed into the crucible and weighed (W2). The sample contained in the crucible was placed in muffle furnace set at 550°C. After ashing the crucible containing the ash was allowed to cool and then weighed (W3).

Calculations:

$$\text{Percentage Ash} = \frac{Weight\,of\,ash}{Weight\,of\,sample} X 100 = \frac{W3-W1}{W2-W1} X 100$$

Where: W1= Weight of crucible.
            W2= Weight of crucible + sample (before ashing).
            W3= Weight of crucible + ash (after ashing).
(Appendix 2).

3.5. 2. Moisture content determination

A clean dry crucible was weighed as (W1), and 5g of the sample material was placed on it, and then weighed, this is W2. The dish was then placed in an oven at 120°C for 3 hours. The dish was removed from the oven, cooled in desiccator for 30 minutes and finally weighed (W3).

Calculations:

$$\text{Percent moisture content} = \frac{Loss\ of\ weight\ in\ drying}{Weight\ of\ sample\ taken} = \frac{W2-W3}{W2-W1} X100$$

Where,   W1= Weight of empty Petri dish.
            W2= Weight of Petri dish + sample.
            W3= Weight of Petri dish + dried sample.
(Appendix 3).

3.5. 3. Lipid content determination

Three gram of the sample was weighed (W1) in to a folded fat free filter paper. This was weighed (W2). The filter paper was carefully placed in an extraction thimble. About 300ml of petroleum ether was added and the apparatus was connected. The extraction was then carried out for three hours using heating mantle, with continuous flow of water in the condenser. The sample was then removed; air dried and then placed in an oven at 80°C until constant weight was obtained (W3). Clean flask was obtained and weighed (W4). After the distillation, the fat containing, petroleum ether was poured into the flask, which was placed in an oven for the petroleum ether to evaporate. After evaporation, the flask containing the fat was weighed (W5).

Calculations:

$$\text{Percent(\%) lipid content} = \frac{W5-W4}{W1} X100$$

Where: W1= Weight of the sample.
            W2= Weight of sample + filter paper.

21

W3= Weight of W2 after extraction.

W4= Weight of empty flask before extraction.

W5= Weight of flask after extraction.

(Appendix 4).

3.5. 4. Protein determination

The sample was weighed (0.15g), and transferred into Kjeldahl digestion flask. This was followed by the addition of 0.8g of a catalyst (0.7g Sodium sulphate, 0.06g copper sulphate and 0.04g mercury II oxide red). Also 2ml of concentrated sulphuric acid H_2SO_4 was added. The mixture was heated on a heating mantle at an inclined position for one hour until the liquid became clear. The digest was then cooled and made alkaline with 15ml 40% Sodium hydroxide (NaOH). The digest was then transferred into steamed out apparatus with minimum volume of water. The ammonia was distilled into 10ml 2% boric acid solution with five drops of methyl red indicator, for 15 minutes. The distilled ammonia was then titrated with 0.02M hydrochloric acid.

Calculations.

1ml of 0.02M HCl =0.00056gN

Xml of 0.02M HCl=YgN

Crude protein content = YgN x factor (6.25) =Pg

Finally % Protein = $\dfrac{Pg \, x \, 100}{0.15g}$

(Appendix 5).

3.6. 0. EXTRACTION OF WATER AND ETHANOLIC FRACTIONS, FROM SPICE INGREDIENTS (PEPPER, GINGER, CLOVES AND WEST AFRICAN BLACK PEPPER).

Three hundred grams (300g) of the spice ingredient powder was suspended in water, and in 99% ethanol in separate flasks. The suspensions were kept at room temperature and left 14 days with regular shacking. The suspensions were then filtered, and the solvents were removed (both water and ethanol) by evaporation to dryness at room temperature (Fatope *et., al.* 1993). This gave the extract used in preparing the stock solution for further analysis.

3.6. 1. PHYTOCHEMICAL SCREENING OF THE SPICE EXTRACTS
3.6.1. 1. Test for alkaloids
This was carried out according to the method described by Ciulci, (1994). To 1.0ml of each extract in two separate test tubes, 2-3 drops of Dragendoff's and Meyer's reagents were added. An orange red precipitate/turbidity with Dragendoff's reagent, or white precipitate with Meyer's reagent would indicate the presence of alkaloids.

3.6.1. 2. Test for flavonoids
This was carried out according to the method of Sofowora, (1993). A piece of magnesium ribbon was added to 4mg/ml of each extract. This was followed by the addition of concentrated hydrochloric acid (HCL), drop wise. Crimson to magenta colour indicated the presence of flavonoids.

3.6.1. 3. Test for saponins.
This was carried out according to the method of Brain and Turner, (1975). Half gram of each extract was placed in a test tube and then 0.5ml of distilled water was added. The tube was then shaken vigorously. A persistent froth that lasted for at least 15mins indicated the presence of saponins.

3.6.1. 4. Test for reducing sugars
This was carried out according to the method of Brain and Turner, (1975). One ml of stock solution of each extract was diluted with 2ml of distilled water, followed by the addition of Fehling's solution (A+B) and the mixtures warmed. Brick red precipitate at the bottom of the test tubes indicated the presence of reducing sugars.

3.6.1. 5. Test for steroids
This was carried out according to the method described by Ciulci, (1994). Two grams of each extract was evaporated to dryness. The residues were dissolved in acetic anhydride, and chloroform was then added. Concentrated sulphuric acid was then added by the side of the test tube. A brown ring at the interphase of the two liquids and the appearance of violet colour in the supernatant layer indicated the presence of steroids.

3.6.1. 6. Test for tannins

This was carried out according to the method described by Ciulci, (1994). The extracts were diluted with distilled water and 2-3 drops of 5% ferric chloride solution was added. A green-black or blue-black colouration indicated the presence of tannins.

3.7. BIOASSAY

3.7. 1. Preparation of extracts impregnated paper discs.

Two grams (2g) of each of the extracts were dissolved in 2ml of appropriate diluent (water for water extract and dimethyl sulpoxide (DMSO) for ethanolic extract), to yield 1.0g/ml (1,000,000μg) solution. This was labeled as the stock solution. From the stock solution 0.1ml was transferred in to a bijou bottle containing 0.9ml diluent, to effect 10 times dilution this gave a concentration of (100000μg/ml). Subsequently 0.1ml was transferred in to another bottle containing 0.9ml diluent which gave a concentration of 10000μg/ml and this was further diluted to yield 1000μg/ml, 100μg/ml, 10.0μg/ml, and 1.0μg/ml on pro-rata basis. A 100 discs, 6.0mm diameter of whatman No. 1 filter paper were impregnated with the extracts to arrive at 1000, 100, 10 1.0μg/disc. Greater disc potencies of 2000, and 3000μg/disc were prepared and stored in refrigerator before use (Deeni and Hussein, 1991)

3.7. 2. Growth Media

Plates of Mueller Hinton agar (Oxoid) were prepared according to the manufacturer's guide. Excess moisture was removed by drying in agar dryer for 15 minutes.

3.7. 3. Preparation of Turbidity Standard

One per cent (1% v/v) solution of sulphuric acid was prepared by adding 1ml of concentrated (H_2SO_4) into 99ml of water. One per cent (1%) weight per volume solution of barium chloride was also prepared by dissolving 0.5g of dehydrated barium chloride in 50ml distilled water. Then 0.6ml of the barium chloride solution was combined with 99.4ml of sulphuric acid solution to yield 1.0% w/v barium sulphate suspension. The turbid solution formed was transferred into a test tube as the standard for comparison (Cheesebrough 2000).

3.7. 4. Standardization of inoculum

Using inoculation loop, enough material from an overnight culture of test organism that is *E. coli, Salmonella* and *Staphylococcus aureus* (isolated from meat) was transferred into a tube containing about 2.0 ml normal saline, until the turbidity of the suspension matched the turbidity of the standard (1% barium sulphate) (Cheesebrough 2000).

3.7. 5. Sensitivity testing of the isolates to the extracts

Two loopfuls of the standard inoculum were evenly streaked on to the plates in duplicates. Discs of different concentrations as well as the control discs (impregnated with only diluent) were placed firmly on the surface of the medium (Mueller Hinton agar) by means of sterile syringe needle at about 40mm apart. The plates were incubated at 37°C for 24 hours. Diameters of the zones of inhibitions were measured with ruler and the mean recorded in the nearest.

3.7.6. Sensitivity testing of the isolates against the commercially produced anitibiotic disks (multi disc)

Two loopfuls of the standard inoculum were evenly streaked on to the plates in duplicates. The multi discs were placed firmly on the surface of the medium (Mueller Hinton agar) by means of sterile forceps. The plates were incubated at 37°C for 24 hours. Diameters of the zone of inhibitions were measured with mm rule and the mean recorded in the nearest mm.

4.00 RESULTS

4.1. 1. Microbiological analysis from abattoir.

The result of the microbiological analysis from abattoir is presented in Tables 4.1 and 4.2 below.

In Table 4.1, the results of the microbiological analyses of swabs from the abattoir are presented. In each case 40 swabs were used. The aerobic plate counts and the fungal counts are presented as log of mean colony forming unit per centimeter square (cfu/cm^2). Out of the 40 batches of swabs from the floor, the tables, the knives and the butcher's hands, 40, 40, 19 and 21 respectively yielded E coli.

In Table 4.2, the result of raw meat and water analysis from the abattoir is presented, The aerobic plate counts, and the fungal counts are presented as log of mean colony forming unit per gram (cfu/g), Coliform count of 180 MPN of coliforms/ml, and E. coli was detected.

In Table 4.2, the number of meat samples from abattoir were 40, the counts are presented as log mean cfu/g and greater than 2400 most probable number (MPN) of coliforms. E coli were isolated from 21 samples, Salmonella spp 20 and Clostridium perfringens 22.

Table 4.1: Microbiological load of swabs from abattoir.

Source of samples n=40	APC Log of Mean cfu/cm^2	FC Log of mean cfu/cm^2	Number of samples with *E. coli*
Floor	6.3502	4.0334	40
Tables	6.1072	4.0410	40
Knives	4.1003	3.3711	19
Hands	4.1038	3.0828	21

Key:

APC = Aerobic plate count.

FC =Fungal count.

cfu/cm^2 = colony forming unit per centimeter square

Table 4.2: Microbial load of raw meat and water from the abattoir.

Source of Sample	Microbial count log Mean (cfu/g)		MPN per g	Identified microorganisms			
	APC	FC	CC	*E. coli*	*E.coli* O157:H7	Salm	C.perf
Raw meat	9.6222	6.5977	>2,400	21	00	20	22
*Water	5.0899	3.3444	180	+	-	-	-

* Counts for water are per ml.

Key:

APC = Aerobic plate count.

SC = Staphylococcal count.

FC =Fungal count.

CC = Coliform count.

cfu/g = colony forming unit per gram.

4.1. 2. Microbiological hazard of *kilishi.*

The results of the microbiological analyses of *kilishi* are presented in Tables 4.3 and 4.4.

In Table 4.3, counts from knives, boxes and hands of *kilishi* processors are presented as natural log of mean cfu/cm^2. Then *E. coli* was isolated from 5 knives and 2 hands.

In Table 4.4, the counts (APC, SC and FC) of the raw meat, spices and roasted meat during *kilishi* production are presented as log mean cfu/g. The coliform counts ranged from 21 in roasted product to >2400 MPN of coliforms/gram of raw meat. *E. coli, Salmonella* and *Clostridium perfringens.* Compared with the other products there was significant difference only with balangu (P≤0.05).

4.1. 3. Microbiological hazard of *dambun nama.*

The results of the microbiological analyses of *dambun nama* are presented in Tables 4.3 and 4.5

In Table 4.3, counts from knives, boxes and hands of *kilishi* processors are presented as natural log of mean cfu/cm^2. Then *E. coli* was isolated from 5 knives and 2 hands.

In Table 4.5, the counts (APC, SC and FC) of the raw meat and the finished product during *dambun nama* production are presented as log mean cfu/g. All the counts are high in the raw meat, low in the finshed product and high in the products held for some time during retail. Neither *E. coli* nor *Salmonella spp* was isolated from the finshed products but two samples had *Clostridium perfringens.* No significant difference between the counts of dambun nama and those of kilishi and tsire at (P≤0.05).

Table 4.3: Microbial loads of swabs from *kilishi* and dambun nama processors.

Source of sample n=20	APC Log of mean cfu/cm^2	FC Log of mean cfu/cm^2	Number of samples with *E. coli*
Knives	4.1239	3.4914	5
Box	2.6532	<30	-
Hands	2.4914	<30	2

Key:

APC = Aerobic plate count.

FC =Fungal count.

cfu/cm^2 = colony forming unit per centimeter square

<30 = Less than 3 X 10^2 (colony count).

Table 4.4: Microbial load of *kilishi*.

Meat sample n=20	Microbial count Log of mean cfu/g		Coliform count	Identified microorganisms			
	APC	FC	CC MPN/g	*E. coli*	*Salm.*	*E coli* O157:H7	*C. perf.*
Raw	9.7924	7.3503	>2400	10	06	00	09
Spice	9.4713	5.0170	>2400	01	01	00	04
Spiced and sun dried	8.6335	5.3181	>2400	20	02	00	14
Roasted	5.4914	4.9494	21-2400	00	00	00	01
Roasted and held	7.0531	5.0828	21->2400	03	01	00	04

Key:

APC = Aerobic plate count.

SC = Staphylococcal count.

FC =Fungal count.

CC = Coliform count.

cfu/g = colony forming unit per gram.

Table 4.5: Microbial load of _Dambun nama._

Meat sample n=20	Microbial count Log of mean cfu/g		CC MPN ml^{-1}	Identified microorganisms			
	APC	FC		_E. coli_	_Salm._	_E. coli_ O157:H7	_C. perf._
Raw	9.7924	7.3502	>2400	10	06	-	09
Cooked	<30	<30	04	-	-	-	-
Fried	<30	<30	<3	-	-	-	-
Fried and held	7.1790	5.1761	240->2400	-	-	-	2

Key:

APC = Aerobic plate count.

SC = Staphylococcal count.

FC =Fungal count.

CC = Coliform count.

MPN=Most Probable Number

cfu/g = colony forming unit per gram.

4.20. Hand culture experiment.

The result of the hand culture experiment was presented in fig 4.1. Following incubation, the plate touched with unwashed hand yielded heavy growth of bacterial colonies at the points of contact between the fingers and the surface of the medium. The plate touched with washed hand remained sterile after 24 hours of incubation.

4.30. Identification of moulds

The various products yielded a variety of fungal species, which included *Aspergillus, Penicillium, mucor* and *Trichophyton.*

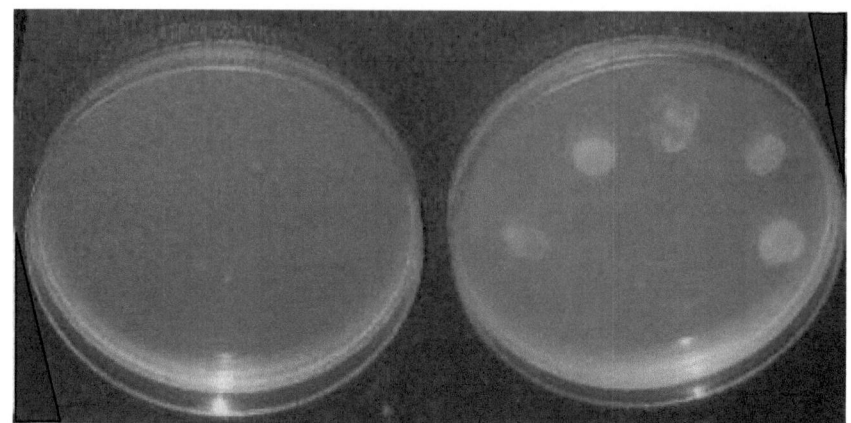

Plate. 4.1. Plates showing cultures from washed hand (left) and unwashed hand (right).

Table 4.6. Biochemical characterization of the isolates.

Tests					KIA		Inference
Indole	MR	coagulase	catalase	oxidase	butt	slope	
+	+		+	-	Y	Y	*E. coli*
		+	+				*Staph. aureus*
-	+		+	-	Y	R/H$_2$S	*Salmonella spp*
-			-	-			*C. perfringens*

Key:

MR   =   Methyl red test.

KIA   =   Kligler iron agar test.

Y =   Yellow.

R =   Red.

+ =   Positive.

- =   Negative.

Table 4.7. Temperature and pH measurements

Product	Temp. of processing (°C)	Time (mins.)	pH	Temp of holding
Kilishi	70	3-5	6.33	25-30
Dambun nama	95	30-60	6.21	25-30

4.40. Proximate composition of the two meat products.

The result of the proximate analysis of the different products is as presented in Table 4.8. *kilishi* and *dambun nama* had percent moisture contents of 4.60 and 8.60, ash of 9.80 and 5.80 respectively, fat of 17.33 and 34.33 and protein of 13.25 and 14.50 respectively.

Table 4.8: Proximate composition of the different meat products.

Sample	Percent moisture	Percent ash	Percent fat	Percent protein	Percent carbohydrate
Kilishi	4.60	9.80	17.33	50.00	18.27
Dambun nama	8.60	5.80	34.33	44.00	7.27

4.5. Physical characteristics and phytochemical composition of the spices.

Results of the physical characteristics and phytochemical analysis of the spice ingredients are as presented in Tables 4.9 and 4.10 below. In Table 4.9, the different extracts have colours ranging from red, brown to dark brown, some with oily and some gummy texture, while all have a spicy smell. In Table 4.10, all the extracts except water extract of West African black pepper contained alkaloids. Flavonoids and reducing sugars were detected in all the extracts. Saponins were detected in all the extracts except ethanolic extracts of West African black pepper. Steroids were detected in water extracts of pepper, ethanolic and water extract of ginger, and water extract of cloves, while in the others no steroids was detected. Tannins were detected in pepper water extract, Ginger water extract, cloves ethanolic and water extract then West African black pepper water extract.

4.5. 1. Bioactivity of the extracts from spices on *E. coli*, *S. aureus* and *Salmonella spp* isolated from the meat products.

The results for the bioactivity of the extracts on the test organisms is as presented in Tables 4.11 and 4.12. In Table 4.11, ethanolic extract of ginger was active against *Salmonella spp* at disc concentrations of 100µg, 1000µg, 2000µg and 3000µg, with zone diameters of 8mm, 19mm, 26mm and 30mm respectively. Ethanolic extract of cloves was active against *Salmonella spp* at disc concentrations of 10µg, 100µg, 1000µg, 2000µg and 3000µg, with zone diameters of 13mm, 24mm, 25mm, 26mm and 26mm respectively. Also the ethanolic extract of West African black pepper was active against *Salmonella spp* at disc concentrations of 10µg, 100µg, 1000µg, 2000µg and 3000µg, with zone diameters of 12mm, 14mm, 20mm, 22mm, and 23mm respectively.

In Table 4.12, ethanolic extract of pepper was active against *S. aureus* at disc concentrations of 10µg, 100µg, 1000µg, 2000µg and 3000µg, with zone diameters of 12mm, 11mm, 16mm, 22mm, and 34mm respectively. Ethanolic extract of cloves was also active against *S. aureus* at disc concentrations of 10µg, 100µg, 1000µg, 2000µg/disc and 3000µg, with zone diameters of 12mm, 13mm, 13mm, 17mm and 22mm respectively.

In the extracts from mixture of all the spices was active only against *Salmonella* and *S. aureus* at the high concentration of 8000µg with zone diameters of 12mm and 14mm respectively.

39

4.5. 2. Sensitivity pattern of *E. coli*, *Salmonella spp* and *S. aureus* to multiple susceptibility disc.

This is presented in Tables 4.13 and 4.14. In Table 4.13, *E. coli* was sensitive to streptomycin with zone diameters of 22mm, ampicilin 24mm, tarivid 24mm, nalidixic acid 14mm, peflacin 20mm, gantamicin 20mm, augumentin 24mm, ciprofloxacin 26mm, septrin 22mm and ceporex 20mm. *Salmonella spp* was sensitive to streptomycin with zone diameter of 22mm, ampicilin10mm, gantamycin 08mm, augumentin 16mm, ciprofloxacin15mm, septrin13mm.

In table 4.14, *S. aureus* was sensitive to streptomycin, erythromycin, ampiclox, zinnacep, pefloxacin, gentamycin, amoxillin, ciprofloxacin, septrin and rocephin with zone diameters of 24mm, 26mm, 24mm, 22mm, 24mm, 26mm, 24mm, 24mm, 26mm and 5mm.

Table 4.9: Some physical characteristics of the extracts from spices.

Extract	Colour	Texture	Solubility	Smell
Pepper ethanol	Red	Oily	DMSO	Spicy
Pepper water	Dark brown	Oily	Water	Spicy
Ginger ethanol	Brown	Gummy	DMSO	Spicy
Ginger water	Dark brown	Gummy	Water	Spicy
Cloves ethanol	Brown	Gummy	DMSO	Spicy
Cloves water	Dark brown	Gummy	Water	Spicy
WABP ethanol	Brown	Oily	DMSO	Spicy

Key :

WABP = West African black pepper.

Table 4.10: Phytochemical composition of spices used in meat processing.

Extract	Alkaloids	Flavonoids	Safonins	Red. sugars	steroids	Tanins
Pepper ethanol	+	+	+	+	-	-
Pepper water	+	+	+	+	+	+
Ginger ethanol	+	+	+	+	+	-
Ginger water	+	+	+	+	+	+
Cloves ethanol	+	+	+	+	-	+
Cloves water	+	+	+	+	+	+
WABP ethanol	+	+	-	+	-	-
WABP water	-	+	+	+	-	+

Key :

WABP =   West African black pepper.

+   = detected.

-    = not detected

Table 4.11: Bioactivity of spice extracts from spice, on *Salmonella spp* isolated from meat.

Extract	Diameter of zone of inhibition (mm)				
	10μg	100 μg	1000 μg	2000 μg	3000 μg
Pepper ethanol	00	00	00	00	00
Pepper water	00	00	00	00	00
Ginger ethanol	00	08	19	26	30
Ginger water	00	00	00	00	00
Cloves ethanol	13	24	25	26	26
Cloves water	00	00	00	00	08
WABP ethanol	12	14	20	22	23
WABP water	00	00	00	00	00

Key :

WABP = West African black pepper.

μg = microgram

Table 4.12: Bioactivity of extracts from spice, on *Staph. aureus* isolated from meat.

Extract	Diameter of zone of inhibition (mm)				
	10µg	100 µg	1000 µg	2000 µg	3000 µg
Pepper ethanol	12	11	16	22	34
Pepper water	00	00	00	00	00
Ginger ethanol	00	00	00	00	00
Ginger water	00	00	00	00	00
Cloves ethanol	12	13	13	17	22
Cloves water	00	00	00	00	00
WABP ethanol	00	00	00	00	00
WABP water	00	00	00	00	00

Key   :

WABP   =   West African black pepper.

µg   =   microgram

Table 4.13: Sensitivity pattern of *E. coli.* and *Salmonella spp* isolated from meat.

Test organism	Diameter of zone of inhibition (mm)									
	S	PN	OFX	NA	PEF	CN	AU	CPX	SXT	CEP
E. coli	22	24	24	14	20	20	24	26	22	20
Salmonella	22	10	00	00	00	08	16	15	13	00

Key:

S   = Streptomycin.

PN= Ampicilin.

OFX   = Tarivid.

NA   = Nalidixic acid.

PEF   = Peflacine.

CN   = Gentamycin.

AU   = Augumentin.

CPX   = Ciprofloxacin.

SXT   = Septrin.

CEP   = Ceporex.

Table 4.14: Sensitivity pattern of *Staph. aureus* isolated fro meat.

Test organism	Diameter of zone of inhibition (mm)									
	S	E	APX	Z	PEF	CN	AM	CPX	SXT	R
S. aureus	24	26	24	22	24	26	24	24	26	25

Key:

S	= Streptomycin.
E	= Erythromycin.
APX	= Ampiclox.
Z	= Zinnacep.
PEF	= Pefloxacin.
CN	= Gentamycin.
AM	= Amoxillin.
CPX	= Ciprofloxacin.
SXT	= Septrin.
R	= Rocephin.

4.6. DETERMINATION OF CRITICAL CONTROL POINTS

Critical control point is a step at which control can be applied, and the application of the control is necessary to eliminate food safety hazard, or reduce it to an acceptable level. Based on the results of the microbiological analysis obtained (tables 4.1, 4.2, 4.3, and 4.4) and using the critical control point decision tree (fig. 2.1), the CCP's in the production of the various products were established.

4.6.1. CCP in *kilishi* production.

Based on the results in tables 4.1, 4.2, 4.3, and 4.5 as well as the CCP decision tree, the CCP in *kilishi* are the raw meat, the spice, drying, roasting then holding and serving to consumers (fig. 4.1).

4.6. 2. CCP in *dambun nama* production.

Based on the results in tables 4.1, 4.2, 4.3, and 4.4, as well as the CCP decision tree, the CCPs in *dambun nama* production are, the raw meat and then holding for retail (fig. 4.2).

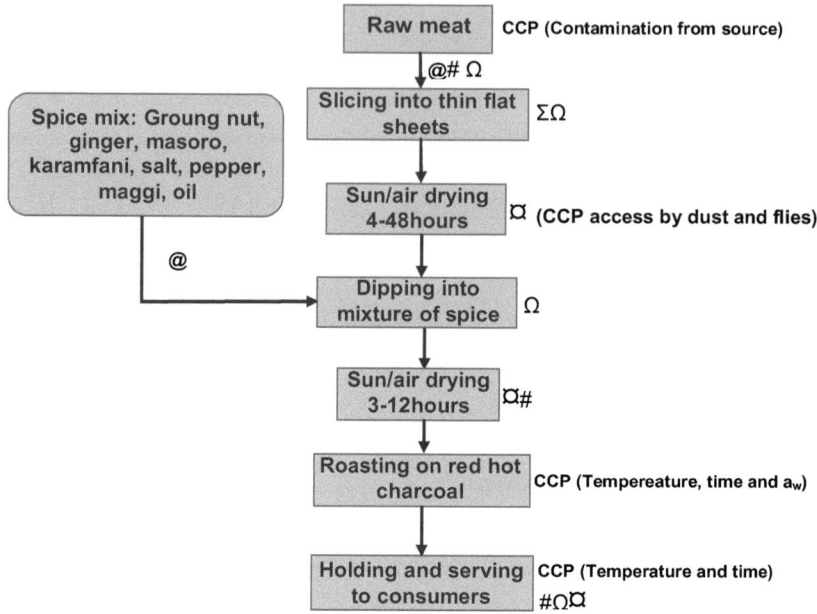

Fig. 4.1: Flow diagram of *kilishi* Production with the Identified Critical Control Points.

Key:

@ = Initial contamination from source.

Ω = Cross contamination

= Multiplication of pathogens possible.

Σ = Contamination of cutleries and utensils

¤ = Contamination from air and dust.

CCP = critical control point.

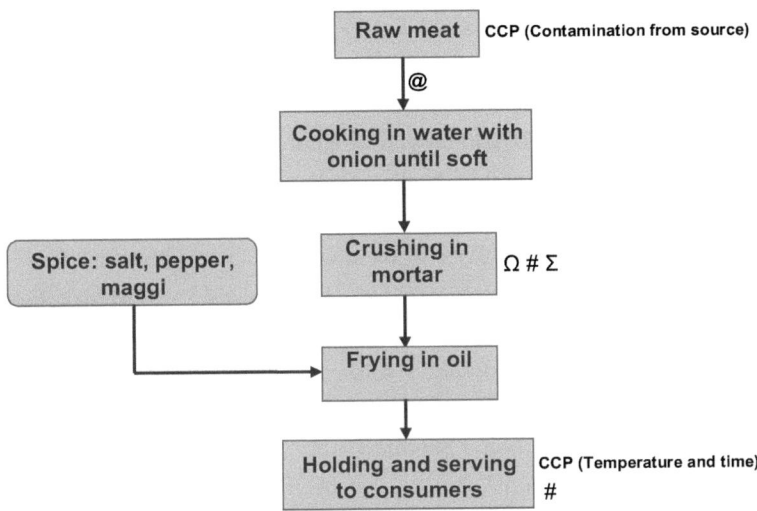

Fig. 4.2: Flow diagram of *dambun nama* Production with the Identified Critical Control Points.

Key:

@	=	Initial contamination from source.
Ω	=	Cross contamination
#	=	Multiplication of pathogens possible.
Σ	=	Contamination of cutleries and utensils
¤	=	Contamination from air and dust.
CCP	=	critical control point.

CHAPTER FIVE

5.00 DISCUSSION

5. 1. Microbiological hazards associated with of the meat products

With respect to the microbiological hazard of meat products in Kano, the present study shows that right from the abattoir there is inadequate hygienic practice and a lot of chances for cross contamination. This is shown in the presence of high bacterial load and indicator organisms on the floor of the abattoir, tables, knives as well as the hands of the butchers. There was also the use of nonpotable water, which had a coliform MPN index of 180. This level is higher than the level provided by FAO (1979), for untreated water which should not have a coliform MPN index of more than 20. According to Umoh, (2001), one of the factors that may contribute to the contamination of meat is the use of non potable water.

The microbiological analysis of the raw meat from the abattoir showed the presence of high counts. These values are actually higher than the acceptable values provided by FAO standards. According to FAO (1979), mesophylic aerobic bacteria, should not be recovered in a number exceeding 10^7 and *Salmonella spp* should not be recovered.

Another point of deviation from FAO standard in the raw meat is the high coliform count and the presence of *E. coli* in particular, which is an indication of improper and in adequate hygienic practices. Abdullahi *et. al.,* (2004), Shamsuddeen and Yusha'u, (2006) also reported the presence of high bacterial load as well as presence of coagulase positive *Staphylococci* in raw meat samples in Zaria and Kano respectively.

The presence of high bacterial load and of those that are pathogenic or toxigenic is hazardous and may be as a result of many factors. According to Umoh, (2001), the possible sources of contamination include using raw meat contaminated through the slaughter of sick animal and handling by butchers, washing meat with nonpotable water, processing product near sewage or refuse dumps, contamination by flies, environment and spices and use of contaminated equipment/utensils.

In the analysis carried out on *kilishi* and *dambun nama*, there was evidence of chances of cross contamination in the processing. This may have effect on the microbial populations on the knives. According to Dusai and Akwar (1999), inadequate cleansing of equipment, infected handlers, water for washing as well as frequent handling can bring about contamination of meat products.

The microbiological analysis of the meat, spices and spiced meat at various stages of processing of the products revealed the presence of high bacterial and fungal counts, as well as presence of indicator organisms and some that are pathogenic. The raw meat, the spice and the spiced meats before roasting in the case of *kilishi* all have counts (APC $>10^7$ in each case) higher than the acceptable levels provided by FAO (1979) and ICMSF (1978).

The, coliform and fungal counts at all stages before roasting were high in addition to the presence of *E. coli, Salmonella spp* and *Clostridium perfringens.* Abdullahi and others, (2004) also reported the presence of *staphylococci* and coliforms in *kilishi.* Shamsuddeen and Yusha'u (2006) reported the presence of *Staphylococcus, E. coli* and *Salmonella* in raw meat just before processing into *balangu* (another traditional meat product).

The occurrence in the spices of high aerobic mesophilic bacterial count, coliform count, fungal count as well as organisms that are potentially pathogenic is an indication of improper handling. According to Ejeikwu and Ogbonna, (1998), presence of pathogenic microorganisms on spices ingredients is considered as major cause of gastrointestinal disturbance resulting from the consumption of meat products. The counts were very high and unacceptable. According to FAO, (1979), aerobic plate count of spices should not exceed 10^6cfu/g and fungal count should not exceed 10^4cfu/g. The presence of high microbial populations on the spices is in agreement with ICMSF (1986), which reported that, spices commonly carry numbers of bacteria, yeasts, mould spores and sometimes insects. Conditions of handling after harvest often permit extensive contamination and microbial growth although drying with heat somewhat reduces microbial numbers. According to Price and Schweigert, (1971), Frazier and Westhoff (2006), unless spices are treated to reduce their microbial content, they may add high numbers and undesirable kinds of organisms to food in which they are used. So this might be an explanation to the very high microbial loads of the mixtures of spices as observed in this study.

The microbial populations of the meat products immediately after roasting dropped down drastically from what was obtained in raw meats. This is not an unexpected outcome because heating generally destroys and reduces microbial cells (Umoh, 2001). This reduction in the microbial populations after heating is more pronounced in *dambun nama* which had APC of less than 30cfu/g. This might be due to the fact that *dambun nama* is more exposed to heat treatment (95°C) than *kilishi* with 70°C). Despite the reductions in the microbial populations of the roasted meat the values for *kilishi* were slightly higher than the acceptable levels by FAO (1979) and ICMSF (1986), according to whom APC of ready to eat meat products should not be in a number exceeding 10^4. For *dambun nama* the values were within the acceptable limit.

The products showed a remarkable increase in microbial populations during the period of holding for retail. *Kilishi* had APC of 3.10 X 10^5 cfu/g after roasting but increased to 1.13 X 10^7 after holding for one to three days and *dambun nama* had APC of <30 cfu/g which increased to 1.51 x 10^7 cfu/g after holding for one to three days. The possible reason for this increase in the microbial load is the way and manner in which the meat operators work. Sometimes the meat products may be left unprotected exposed to dust. This could introduce microbes in addition to the ones introduced through the butcher's hands (cross contamination). Microbes are introduced to *kilishi* from the air and the hands of the sellers because during retailing, *kilishi* is not cut with knife but rather broken in to pieces with bare hands. For *dambun nama* also the microbes are introduced through the butcher's hands during packaging. This is because normally after frying, the sellers use bare hands to put the minced meat into the retailing plastic containers.

5. 2. Hand culture experiment.

From the result of the hand culture experiment, it was clear that there was contamination on the hands of the meat processors. Washing of hands with clean water and detergent is very effective in eliminating the contaminants since the culture from washed hand yielded no growth after incubation. This means, if meat handlers should be washing their hands before handling there will be a remarkable decrease in the chances of cross contamination.

5.3. Moulds isolation.

Presence of variety of mould species and their high count are points of great concern. This is because, mould like *Aspergillus* has the ability to produce extracellular metabolites called the mycotoxins. Example is the aflatoxins which are a group of highly toxic meatabolites produced by the fungi of the genus *Aspergillus*. This study agrees with the work of Badau *et. al.,* (2001) and Abolama and Egbebi (2007). who reported the presence of *Aspergillus*, *Penicillium* and *Mucor* in meats at Maiduguri and Ado Ekiti respectively.

5.4. Temperature of processing, holding and pH of the products.

It was evident from the results of this study that the processing temperature of the two products (*kilishi* and *dambun nama*) differ. This difference in the processing temperatures may be responsible for the difference in the microbial load of the products immediately after roasting. For instance; *dambun nama* had a relatively low microbial load immediately after frying. This is actually due to the high temperature treatment they undergo during the production (>80°C for over 30mins). In the case of *dambun nama* it even involved two types of high temperature treatment, cooking in water and frying in oil, both at high temperature. High microbial loads were however recorded in *dambun nama* when held for some time before selling. This is as a result of the way the meat is handled during packaging. Some of the sellers used bare hands in dispensing of the products into the plastic containers and also the environment that was unhygienic. According to Igene and Abulu (1983), influence of environmental sanitation on the microbial population is a highly significant factor in the quality of retailed meat.

kilishi is roasted at relatively lower temperatures (70°C) than *dambun nama* (95°C). Bryan *et. al.*, (1980) reported that, product maintained at high temperature, do ensure the killing of vegetative cells and reduction in moisture content. The spices of *kilishi* also contributed a lot to the high microbial load of the product.

In the case of *kilishi*, the product is roasted at about 70°C but for just three to five minutes depending on the strength of the fire and the position of the meat on the roasting wire gauze. The high microbial load might be as a result of short time heat treatment, spices and handling. The high loads could be reduced by improving personal and environmental sanitation, preparation of spices with clean water since the product is roasted for a short period of time. The spices should also be treated to reduce their microbial loads, suggestively using heat treatment.

The pH of the products and holding temperatures might also have a great influence on the microbial populations of the products. The pH values of *kilishi* and *dambun nama* which were found to be 6.33 and 6.21 respectively, and the holding temperatures which ranged between 25 and 40 were all not unfavourable for the survival of the microbes. This is because they are mainly aerobic mesophylic bacteria. According to Robert, (2004), Aerobic mesophylic bacteria grow well at temperature between 20°C and 40°C. They also grow well at pH values near neutrality (Frazier and Westhoff, 2006).

5. 5. Proximate analysis of the products.

One important factor affecting microbial growth on any food is water or the moisture content of the food. Of the two products, *dambun nama* had the higher moisture content (8.60%) than *kilishi* which had (4.60%).

High moisture content favours the growth of many spoilage and pathogenic microbes and this is one of the reasons why meat is considered as an excellent medium to support microbial growth (Adams and Moss 1995). This means, meat has a high water activity. Among the two products, *dambun nama* even though having low counts after frying, would be more prone to spoilage since it has the highest moisture content.

Kilishi had the least moisture content (4.60%) because of three factors. First the meat is sliced in to thin flat sheets, which makes water to evaporate quickly; it is then dried twice during the processing and finally roasted though for a short time (usually 3-5 minutes). This could be the reason why *kilishi* had the lowest moisture content. However the high microbial load might have come from the spices and the raw meat since the product is roasted for just 3-5 minutes at around 70°C, which might not be enough to eliminate contaminating organisms.

The ash content had an inverse relation to the moisture content. *kilishi* which had the least moisture content (4.60%) had the highest ash content (9.80%). So based on this the higher the moisture content, the least the ash content and hence the shorter the shelf life. Conversely, the lower the moisture content the higher the ash contents hence the longer the shelf life of the product. Presence of protein carbohydrate and lipids in the products means presence of degradable substances since the various contaminating organisms possess various proteolytic and lypolytic enzymes.

5.6. Bioactivity testing of spice ingredients.

According to Pruthi, (1980), properties of spices include bactericidal, bacteriostatic, fungistatic, antihelminstic, medicinal and flavouring. From the result of this study it is clear that, not all the spices extracts were active against the test organisms. Even those that were active were of the ethanolic extracts. The water extracts were not active. The mixture of all the extracts of the spices was only slightly active against *Salmonella spp* and *Staphylococcus aureus* at concentration of 8000μg/disc. This signifies that, individually the extracts of the spices are more active against the test organisms than when they were combined. According to Frazier and Westhoff (2006), spices and condiments do not have any marked bacteristatic effect in the concentrations normally used but may help other agents in preventing the growth of microorganisms. They also showed that unless the spices are treated to reduce their microbial content, they may add high numbers and undesirable kinds of microorganisms as observed from this study.

5.7. Critical Control Points Determination

Based on the critical control point decision tree a stage is a critical control point if it involves a hazard of sufficient likelihood of occurrence and severity to warrant its control, control measure exist, and the control at the step is necessary to prevent, eliminate or reduce the risk of the hazard or reduce it to an acceptable level (NACMCF, 1997).

For both the products, the raw meat and spices had microbial loads higher than the acceptable levels provided by FAO, 1979 and ICMCF, 1978). There were also high fungal and coliform counts. *E. coli, Salmonella spp* and *C. perfringens* were also isolated in many of the raw meat samples. All these are hazards of sufficient likelihood of occurrence and severity to warrant their control. The control measures exist and the control measures that could be applied include; observing strict hygienic practices during slaughtering, skinning, cutting and distribution of the raw meat to retailers. Attention should also be given to temperature and time of holding so that they should not be favourable for the proliferation of the contaminating microbes. Therefore the raw meat and the spices were CCPs. According to Adam and Moss (1995), a raw material itself could be a critical control point.

In the production of *kilishi*, the stages identified as CCPs are the raw meat and, drying, roasting and holding for retail. These stages are CCPs because they all involved hazards that needed to be controlled, the control measures exist and application of the control measures is necessary in order to eliminate or reduce the hazards to an acceptable level. According to Abdullahi *et. al.,* (2004), the main CCPs in *kilishi* processing was the drying stage.

In the production of *dambun nama* the main CCP apart from the raw meat itself was the last stage, that is the period of holding during retail. This is because of the hazard (high counts and presence of hazardous organisms) that exist at that stage, and there is need for the application of the control measure in order to eliminate the hazard or reduce it to an acceptable level. The possible control measure is protecting the fried meat from flies and dust, and also using a clean scoop or spoon during the dispensing of the meat in to clean plastic containers, because some of the meat handlers use bare hands for dispensing of the meat.

5.8. Conclusion

Based on the findings from this study, hazards exist at various stages in the production of the meat products, these hazards include high bacterial and fungal counts, presence of indicator organisms as well as organisms that are potentially pathogenic. Stages including the raw materials (raw meat and spice) are CCPs. Microbiological hazards could be eliminated or reduced to an acceptable level if strict hygienic measures are taken. Ethanolic extracts from spices were more active individually, but water extracts were not active on the test organisms. Spices used in the meat products increased their bacterial load rather than reduced it. There were evidences of cross contamination in the production line because of high loads and presence of indicator organisms and organisms that are potentially pathogenic.

Occurrence of hazards is the result of unhygienic practices right from the abattoir, use of non potable water for meat washing; cross contamination from butchers hands, knives, tables, and other utensils, allowing flies and dust to have access the meat.

5.9. Recommendations

- ❖ There should be provision of potable water at the abattoir in order to ensure adequate cleanliness of the floor, tables, meat and knives of the butchers at the abattoir.
- ❖ Butchers should avoid mixing of carcass with intestinal contents which are the major source of contaminating organisms.
- ❖ All meat operators and handlers should exercise meat operators and handlers should exercise personal and environmental hygiene at all stages to avoid cross contamination.
- ❖ Flies and dust should be denied access to meat and meat products.
- ❖ Spices should be treated to reduce their microbial loads, and potable water should always be used to in their preparation.
- ❖ During the packaging of dambun nama, clean spoon or scoop should be used rather than using bare hand.
- ❖ Adequate heat should be applied while roasting any type of meat product.
- ❖ Researchers should always educate meat handlers and operators based on their research findings.
- ❖ Government should also come in by putting in place, policies that will ensure compliance with standard hygienic practices in meat handling and operations.

REFERENCES

Abdullahi, I. O., Umoh, V. J., Ameh, J. B. and Galadima M. (2004): Hazards associated with Kilishi preparation in Zaria, Nigeria. *Nigerian Journal of Microbiology.* Vol. **18** (1-2): 339-345).

Aboloma R.I. and Egbebi A. O. (2007). Microbiological quality of some dried meat and fish samples sold in Ado Ekiti, Ekiti State, Nigeria. *BEST JOURNAL* 4(2):158-161.

Adam M. R. and Moss M. O. (1995): *Food Microbiology.* Royal society of chemistry publishers Pp115-116.

Atlas, R. M. (1997). Principles of Microbiology second edition C. Brown publishers. Pp 802-803.

Badau, M. H. Adeniran, A. M. and Nkama, I. (2001). Fungi associated with various fresh meat sold in Maiduguri market, Nigeria. *Science Forum: J. Pure andAppl. Sci.* 4(2)255-262.

Bourquim, R. G. (1980). Industrial process for achieving long term preservation of meat food matter. *French patent Appl.* 451- 715'

Brain, K. R. and Turner, T. D. (1975). The practical evaluation of phytopharmaceuticals. Wright Scientechica, *Bristol*: 57-58.

Broidy, B. A. and Kaminsky, Z. C. (1978). Enteric and urinary tract infections in: Broidy, B. A. (ed). *Microbiology and infectious diseases.* McGraw Hill Book Company, New York. P 404-406.

Bryan, F. L., Bartelson, C. A., Christoperson, N. (1980). Hazard analysis in reference to Bacillus cereus of boiled ricein Cantonese-style restaurants. Journal of Food Protection 44: 500-512.

Bryan, F. L. (1992). Hazard Analysis Critical Control Point Evaluations, a guide to identifying hazards and assessing risks associated with food preparation and storage. W. H. O. Geneva.

Canadian Food Inspection Agency (CFIA) (1997). HACCP generic model. Dried meat (beef jerky) Canadian Food Inspec. Agency.

Cappuccino J. G. and Sherman N. (2002). Microbiology laboratory manual, sixth edition. Benjamin Cumming publishers. Pp 121-122.

Cheesebrough, M. (2000). District Laboratory practice in Tropical Countries. Part 2. Cambridge University Press. Pp76-100.

Ciulci, I. (1994). Methodology for the analysis of vegetable drugs. Chemical industries branch, Division of industrial operations. UNIDO, Romania: Pp 24, 26 and 67.

Clark, D. (1991). FSIS Studies detecton of food safety hazards. *FSIS Food Saf Rev* 4: Summer.

Claude, M. and Maurice, M. (1979). Moulds toxins and food. Wiley Inter Science Publishers. Pp 16-26.

Cloak, O. A., Duffy, G., Sheridan, J. J., McDowell, D. A. and Blair, I. S., (1999). Development of surface adhession immunofluorescent technique for the rapid detection of Salmonella spp, from meat and poultry. Journal of Applied Microbiology 86, 583-590.

Deeni Y. Y. and Hussein, H. S. N. (1991): Antibiotic sensitivity testing. Journal of ethnopharmacognosy. Elseiver Scientific Publishers, Ireland pp 91-96.

Dincer A. H. and Baysal T., (2004). Decontamination technique of pathogenic bacteria in   meat and poultry. *Critical Reviews in Microbiology.* 30: 197-204.

Dusai, D. H. M. and Akwar, H. T. (1999). Total plate count, isolation and antibiogram of *Escherichia coli* in ready to eat *suya* in Zaria, Nigeria. *Tropical. Veterinary Journal* . **17**, 31-35.

Egan, H. Kirh, R. S. and Sawyer, R. (1981). Pearson's chemical analysis of foods, 8th ed. Churchill Livingstone, Edinburgh, Pp 8-12, 15-29, 192-193.

Ehiri, J. E., Azubike, M. C., Ubbainu, C. N. Anyanwu, B. C., Ibe, K. M. and Ogbonna, M. O. (2001). Critical control points of complementary food preparation and handling in eastern Nigeria. *The International Journal of Public Health* 79(5): 423-433.

Ejeikwu, E. O. and Ogbonna, C. I. C. (1998). Species of microorganisms associated with the ingredients used in the preparation of the Nigerian take away roasted meat (Suya). *Zuma Journal of Pure and Applied Sciences.* 1(1): 7-10.

Erickson, J. P., Stammer, W.J., Haites, M., Mekenna, N. D. and Van Austine, A. C. (1995). An assessment of *E. coli* O157:H7 contamination risk in commercial mayonnaise from pasteurized egg and environment, sources and behaviour, low pit dressing. *Journal of Food Protection* 58: 1059-1064.

Fatope A. O., Ibrahim H. and Takeda Y. (1993): Screening of higher plant reputed as pesticides using the brine shrimp lethality bioassay. *International Journal of pharmacognosy.* 31: 250-256.

Food and Agriculture Organization of the United Nations FAO (1979): Manuals of food quality control 4. Microbiological analysis. D1-D37.

Food Standard Australia New Zealand. (2003). Food poisoning bacteria. Fact shet.

USDA/FSIS (1998) *Microbiology Laboratory Guide book 3rd Edition.*

Forest, J. C. Aberle, E. D., Hedrick, H. B., Judge M. D., and Merkel, R. A. (1975). *Principle of meat science.* W. H. Freman and co. San Francisco USA Pp1-4 and 228-229.

Fawole, M. O. and Oso, B. A. (2001). Laboratory manual of Microbiology. Spectrum Books Ltd. Pp 21, 79-80.

Fox, J. B. and Ackerman, S. A. (1970). Meat chip, a new snack idea. *Food Microbiology* **24**, 34.

Frazier, W. C. and Westthroff, W. C. (2006). Food microbiology 3rd Edition, McGraw Hill Publishing Company Limited New York. Pp163-165, 223-236, 419-543.

Frey, D., Oldfield, R. J. and Bridger, R. C. (1981). A colour atlas of pathogenic fungi. Wolf Medical Publications. Pp 10-70

Gregory, G. S., Richard, B. O., Michael, P. H., and Changapa, M. M. (1994). Detection of *Salmonella* serovars from clinical samples by enrichment broth cultivation PCR procedure. *Journal of Clinical Microbiology.* **32**: 1742-1749.

Holley, R. A. (1985). Beef jerky: Viability of food poisoning microorganisms on jerky during its manufacture and storage. *Journal of Food Protocol.* 48: 100-106.

Igene, J. O. and Abulu, C. (1983). Physical and nutritional quality of tsire-type suya. A popular Nigerian meat product; *Journal of Food Protocol.* **46** 293-296.

Igene, J. O., Farouk, M. M., Akanbi, C. T. (1989). Preliminary studies on the quality characteristics of kilishi. *Nigeria Food Journal* **7**: 29-37.

International Commission on Microbiological Specifications for Foods (ICMSF) (1986). Microorganisms in food sampling for microbiological analysis: Principle and specific application 2nd Edition. Blackwell Scientific Publications Pp139-140; 213-215.

International Commission on Microbiological Specifications for Foods (ICMSF) (1978). Microorganisms in foods vol 2: Sampling for Microbiological analysis: principle and specific application. University of Toronto Press 110-127. International Standards Organizatio, (ISO) (1972). Spices and condiments. *Women culture Finchized Draft Proposal.* Tc-34/Sc-7, 150 Budapest.

Miller, J. D. (1996). Food borne natural carcinogens. Issues and priorities. Key note presentation at the second of the Pan African Environmental mutagen Society. Café Town, South Africa.

National Advisory Committee on Microbiological Criteria for Foods NACMCF (1997). Hazard Analysis and Critical Control Point, Principles and Guide lines. Pp1-13.

Norman, G. M. and Robert, B. G. (2006). Principles of food Sanitation. 5th edition. Springer Publishers, Pp 99-114.

Nottermaus, S., and Hoegenboom-verdegaal, A. (1992). Existing and emerging food born diseases. *International Journal of Food Microbiology* **15**: 197-205.

Okonkwo, T. M., Obanu, Z. A. and Oneuka, N. D. (1994). Quality characteristics, amino acid and fatty acid profiles of some Nigerian traditional hot smoked meat products. Nigeria Food Journal 12: 46-54.

Oranusi, S. U., Umoh, V. J. And Kwaga, J. K. P. (2003). Hazards and Critical Control Points of Kunun-zaki, a non-alcoholic beverage in Northern Nigeria. *Food Microbiology* 20: 127-132.

Patel, J. D., Krishnaswamy, M. A., and Nair, K. K. S. (1976). Biochemical characteristics of some coliforms isolated from spices. *Journal of Food Science Technology. India* **13**: 37-40.

Pitt, J. I. (1996). What are Mycotoxins? *Australian Mycotoxins News Letter.* 7(4). 1.

Price, J. F. and Schweigert B. S. (1971). *The science of meat and meat products.* 2nd Edition. Published by W. H. Freeman and Company Pp 289.

Pruthi, J. S. (1980). Spices and condiments: Chemistry, microbiology, biotechnology. *Adv. Food Res.* Supple 16-17. Academic Press, New York.

Rajworsky, J. K. and Marmer, S. B. (1995). Growth of *E. coli* O157:H7 at fluctuating incubation temperature. *Journal of Food Protection*, **58**: 1307-1313.

Robert W. B. (2004). Microbiology. Benjamin Cummings Publishers Pp748-753.

Saide-Alboronoz, J. J., Konipe, C., Maurano, E. A., Beran, W. G. (1995). Contamination of pork carcass during slaughter, fabrication and chilled storage. *Journal of Food Protection* 58: 993-997.

Shamsuddeen U. and Yusha'u M. (2006). Bacteriological Quality assessment of meat sold around red bricks (Jan bulo) Quarters in Kano. *Biological and Environmental Sciences Journal for the Tropics* 3(4): 95-98.

Sofowora, A. A. (1993). Medicinal Plants and Traditional Medicines in Africa. Spectrum books Ltd., Ibadan Nigeria: 2. Pp81-85.

Umoh, J.U. (2001) : An over view of possible critical control points of ready-to-eat beef products of northern Nigeria. *International Conference on Food and Security*. Conference center Ibadan, Nigeria. Pp109-115.

Umoh, V. J., and Odoba, M. B., (1999). Safety and quality evaluation of street vended foods sold in Zaria, Nigeria. *Food Control* **10**: 9-14.

Uraih N. (2004). Public health food and industrial microbiology. Bubpeco publishers. Pp88-100.

World Health Organization, WHO (1976). *Microbial aspects of food hygiene.* Technical   Report series. 598: 9-101.

APPENDIX 1

Map of Kano showing the sampling sites.

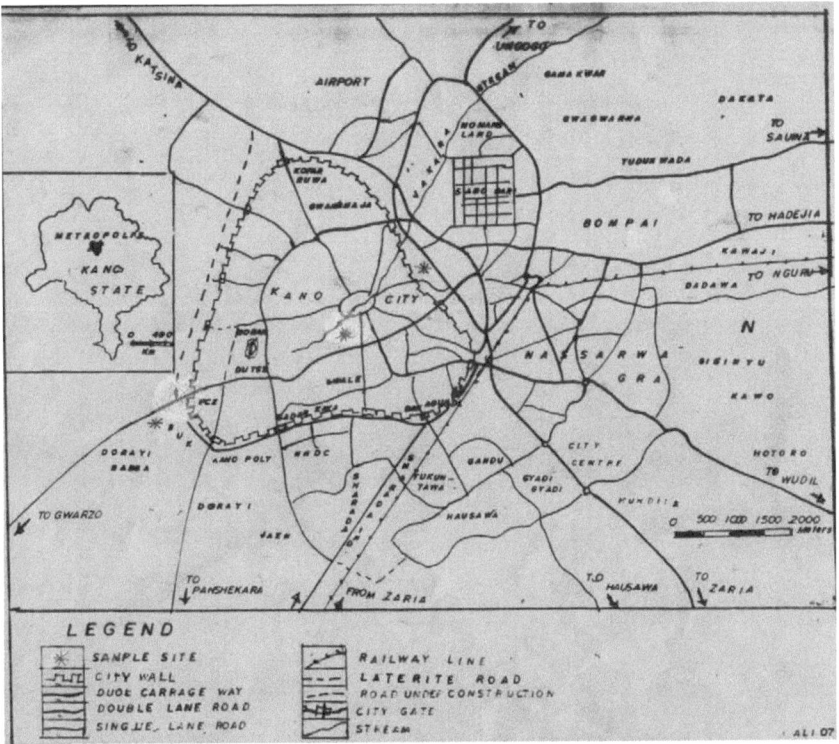

Source: Department of Geography, Bayero University, Kano.

Appendix 2. Data for determination of percent Ash content.

Sample	Weight pf sample		W1	W2	W3
Kilishi	5.00		0.03	25.03	20.52
Dambun nama	5.00		18.35	23.35	18.64

Where: W1= Weight of crucible.

W2= Weight of crucible + sample (before ashing).

W3= Weight of the ash (after ashing).

Appendix 3. Data for determination of percent Moisture content.

Sample	Weight pf sample	W1	W2	W3
Kilishi	5.00	28.26	33.26	33.03
Dambun nama	5.00	40.09	45.09	44.66

Where, W1= Weight of empty Petri dish.

W2= Weight of Petri dish + sample.

W3= Weight of sample taken.

Appendix 4. Data for determination of percent Fat content.

Sample	W1	WFP	W2	W3	W4	W5
Kilishi	3.00	0.63	3.63	3.08	106.06	106.58
Dambun nama	3.00	0.65	3.65	2.47	99.71	100.74

Where: W1 = Weight of the sample.

W2 = Weight of sample + filter paper.

W3 = Weight of W2 after extraction.

W4 = Weight of empty flask before extraction.

W5 = Weight of flask after extraction.

WFP = Weight of filter paper.

Appendix 5. Data for determination of percent Protein content.

Sample	Acid at end point (ml).
Kilishi	21.40
Dambun nama	18.80

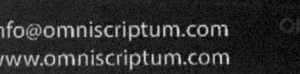